A TRUE BOOK™

OUR UNIVERSE

STARS

Joan Marie Galat

Children's Press®
An Imprint of Scholastic Inc.

Content Consultant
Hsiang Yi Karen Yang, PhD
Assistant Professor
Institute of Astronomy
National Tsing Hua University
Hsinchu, Taiwan R.O.C.
(Former Computational Astrophysicist and Assistant Research Scientist in the
Department of Astronomy at the University of Maryland, College Park)

Dedication: For Grant, the star in my sky.—JMG

Library of Congress Cataloging-in-Publication Data
Names: Galat, Joan Marie, 1963- author.
Title: Stars / Joan Marie Galat.
Other titles: True book.
Description: New York: Children's Press, an imprint of Scholastic Inc., 2021. | Series: A true book | Includes
    index.| Audience: Ages 8-10. | Audience: Grades 4-6. |
Summary: "Book introduces the reader to stars"--Provided by publisher.
Identifiers: LCCN 2020004602 | ISBN 9780531132197 (library binding) | ISBN 9780531132371 (paperback)
Subjects: LCSH: Stars--Juvenile literature.
Classification: LCC QB801.7 .G33 2021 | DDC 523.8--dc23
LC record available at https://lccn.loc.gov/2020004602

Design by Kathleen Petelinsek
Editorial development by Priyanka Lamichhane

Scholastic Inc., 557 Broadway, New York, NY 10012

1 2 3 4 5 6 7 8 9 10 R 30 29 28 27 26 25 24 23 22 21

Safety note! The activity suggested on pages 42 and 43 of this book should be done with
adult supervision. Observe safety and caution at all times. The author and publisher disclaim
all liability for any damage, mishap, or injury that may occur from engaging in the activity
featured in this book.

Front cover: A glowing spiral galaxy in outer space

**Back cover: The Gran Telescopio Canarias, located on the
Canary Islands of Spain, is the largest telescope in the world.**

Find the Truth!

Everything you are about to read is true *except* for one of the sentences on this page.

Which one is **TRUE**?

T or F Stars can fall from the sky.

T or F Every star is born in a nebula.

Find the answers in this book.

Contents

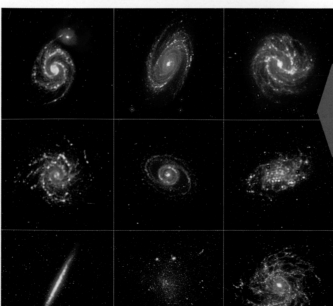

Galaxies have different shapes and are different sizes.

The BIG Truth

The Hubble
Space Telescope

The Eagle
Nebula

Introduction

A Universe of Stars

When the sun goes down, **stars** begin to speckle the darkening sky. You can see thousands just by looking up from a dark location on a clear, moonless night. In some places, stars are so close together, their glow forms a pathway of light in the sky. In others, they are farther apart. **Our universe is home to billions of trillions of stars,** including our sun.

Our sun is the nearest star to Earth. The most **massive stars are 1,500 times bigger than the sun**. Less massive stars are closer to the size of planet Jupiter. **Astronomers**, the scientists who study stars, learn about them by studying the energy of the light stars emit. They use powerful **telescopes** to observe the stars in our universe. And they are making exciting new discoveries all the time!

Many trillions of stars shine
throughout the universe.

Star Traits

The universe contains different types of stars, and each type of star has specific traits. These include a star's size, color, and temperature. Stars are also classified, or grouped, based on their brightness, or how much light they create. Astronomers observe all of these traits in order to gather as much information as they can about these glowing balls of gas.

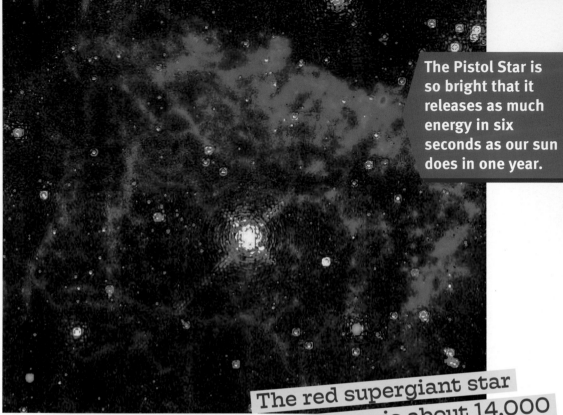

The red supergiant star Betelgeuse is about 14,000 times brighter than our sun!

Star Brightness

Astronomers describe star brightness by comparing **luminosity**. This is the amount of light a star emits, or gives off. The more light a star emits, the more luminous it is. A star's luminosity depends on its size. Large stars have more surface area to release more energy, so they are the most luminous. Stars that are closer to Earth also seem brighter.

Star Size

The easiest way to imagine the size of a star is to compare it to our sun, which is a medium-size star, known as a yellow dwarf. The smallest stars, red dwarfs, are also the most common stars in the universe. They are less than 10 percent the size of the sun. The largest red dwarfs are half the size of the sun. The most massive stars are called giants and supergiants. Giant stars are up to 100 times more massive than the sun. Supergiants can be hundreds of times larger.

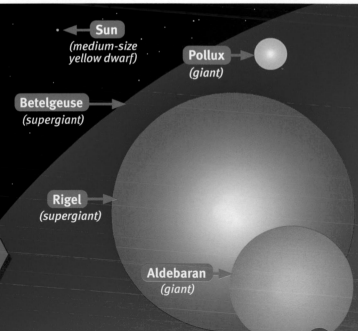

Sun
(medium-size yellow dwarf)

Pollux
(giant)

Betelgeuse
(supergiant)

Rigel
(supergiant)

Many stars are larger than our sun.

Aldebaran
(giant)

Color and Temperature

A star's color depends on its temperature. The hottest stars are blue. Their surface temperature can be 10 times hotter than the sun's. Blue-white stars are the second hottest, and yellow stars are third hottest. Orange stars are cooler, and the coolest stars are red.

Stars range in temperature. Blue stars are the hottest and red stars are the coolest.

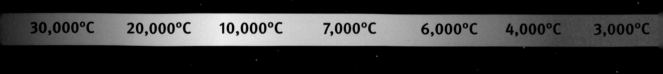

| 30,000°C | 20,000°C | 10,000°C | 7,000°C | 6,000°C | 4,000°C | 3,000°C |

Our Closest Star

The sun is a medium-size star. It is 4.6 billion years old and the closest star to Earth. From 93 million miles (150 million kilometers) away, it warms our planet, providing the heat and light needed for life to exist on Earth. The sun causes all of Earth's weather. Its activity can also disrupt GPS and communication networks.

It takes about 8 minutes for sunlight to reach Earth.

Sun

Earth

The sun is the largest object in our solar system.

Explosions from dying stars release gas and dust into space.

The Orion Nebula, shown in this telescope image, appears as a fuzzy patch when seen with the naked eye. Stars are being born here.

CHAPTER 2

A Star Is Born

Stars go through life cycles marked by different stages. They are born in clouds of gas and dust called **nebulas**. Stars form thanks to **gravity**—a force that draws objects together. Gravity brings dust and gas clouds closer. Then clumps of matter form. The matter grows more and more dense, and gravity causes the matter to collapse. An extremely hot core forms. The resulting object is known as a **protostar**.

Growing Up

A collapsing cloud in a nebula may create hundreds to thousands of protostars. The protostar stage lasts from 100,000 to 10 million years. At the end of this stage, large protostars form a type of star called a main sequence star. Smaller protostars form brown dwarfs. Brown dwarfs are too large to be planets and too small to be stars.

New stars form in pillars of gas and dust in the Eagle Nebula.

We Are Made of Stardust

Six **elements** make up the building blocks of life: sulfur, phosphorus, nitrogen, hydrogen, carbon, and oxygen. They are all part of our bodies and part of life around us. Astronomers have found all six of these elements in the stars in our universe!

Studying stars helps astronomers understand how elements form and how life began on Earth.

Percentage of Elements in the Human Body

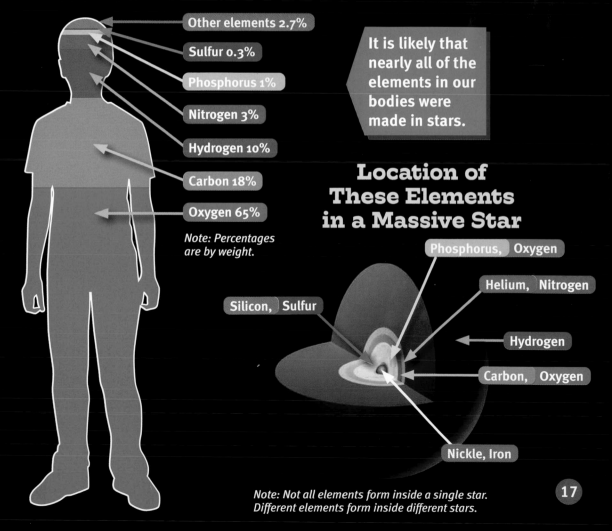

Other elements 2.7%

Sulfur 0.3%

Phosphorus 1%

Nitrogen 3%

Hydrogen 10%

Carbon 18%

Oxygen 65%

Note: Percentages are by weight.

It is likely that nearly all of the elements in our bodies were made in stars.

Location of These Elements in a Massive Star

Phosphorus, Oxygen

Helium, Nitrogen

Silicon, Sulfur

Hydrogen

Carbon, Oxygen

Nickle, Iron

Note: Not all elements form inside a single star. Different elements form inside different stars.

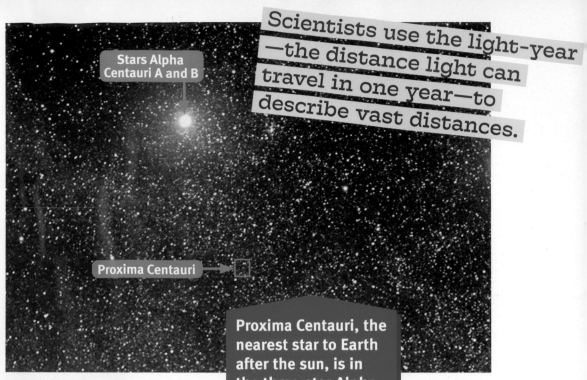

Stars Alpha Centauri A and B

Scientists use the light-year —the distance light can travel in one year—to describe vast distances.

Proxima Centauri

Proxima Centauri, the nearest star to Earth after the sun, is in the three-star Alpha Centauri system.

Star Systems

Some stars, like our sun, are the only star in a region of space. More often, stars are part of star systems. These are groups of two or more stars held together by gravity. Double star systems, with two stars near each other, are called binary pairs. The closest star system to Earth is a triple-star system known as Alpha Centauri. It is 4.4 **light-years** away.

Star Clusters

Some stars are part of a cluster. A star cluster is a group of stars formed from dust and gas in the same region in space. A cluster may include hundreds to thousands of stars held together by gravity. The Pleiades—a cluster with more than 3,000 stars— is found in the constellation Taurus, the bull. Constellations are star groups that appear to form pictures in the night sky. While stars in a cluster are formed in the same region, stars in constellations are formed in different areas in space.

Taurus

Pleiades

There are 88 official constellations across Earth's sky. They form a variety of pictures including mythical creatures, people, and animals.

The Latin word for star is *stella*. You can also find it in the word constellation.

Galaxies

Star clusters contain many stars, but galaxies contain many, many more! A galaxy is an enormous, rotating collection of gas, dust, stars, and their planets—all held together by gravity. Scientists estimate that the universe is home to about 200 billion galaxies. And sometimes galaxies collide! In about five billion years, our Milky Way galaxy and the Andromeda galaxy will crash together. However, individual stars won't smash together because they are so far apart.

Galaxies have different shapes. Some look like spirals, others are egg-shaped, or elliptical. Some don't seem to have much shape at all.

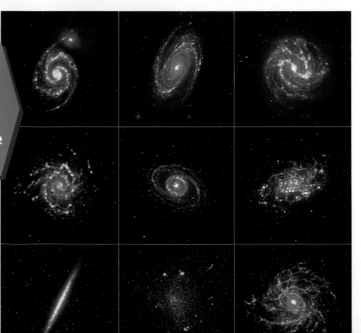

All of the stars you can see from Earth are in the Milky Way galaxy.

The Milky Way

The Milky Way galaxy is shaped like a barred spiral. It stretches 200,000 light-years across the sky and contains more than 200 billion stars. If you go outside on a dark, clear night, you can see that some of these stars form a band of light. The ancient Greeks had a story to explain this trail of light. They said it came from the goddess Hera, who spilled her milk across the sky while feeding Hercules. This myth gave our galaxy its name.

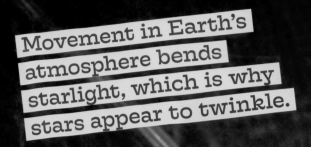
Movement in Earth's atmosphere bends starlight, which is why stars appear to twinkle.

Sirius A →

← Sirius B

The main sequence star Sirius A and the dim white dwarf Sirius B revolve around each other once every 50 years.

The Longest Star Stage

Stars spend the longest part of their life cycles in the phase known as the main sequence stage. Most stars, including the sun, are in this stage of their cycle. During this time, heat and pressure inside the star bring hydrogen atoms together. This process is called **nuclear fusion**. When nuclear fusion happens, atoms join, or fuse, in a star's core. Fusing hydrogen creates helium along with lots of energy in the form of heat and light. This is what makes a star shine.

A Balancing Act

All main sequence stars fuse hydrogen into helium. The reason this stage of a star's life lasts the longest is because the effects of gravity and fusing hydrogen balance each other out and keep the star stable. Nuclear fusion heats up the star's core, creating pressure. That pressure pushes outward. At the same time, the force of gravity pulls inward. The outward and inward forces keep the star stable. A star's stable state is called equilibrium.

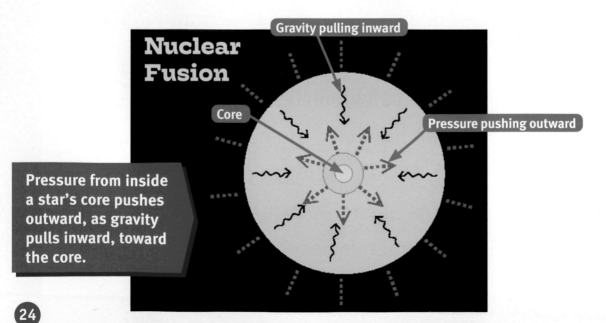

Nuclear Fusion

Gravity pulling inward

Core

Pressure pushing outward

Pressure from inside a star's core pushes outward, as gravity pulls inward, toward the core.

The sun will run out of hydrogen fuel in about five billion years. This will be the end of its main sequence stage.

When Stars Become Unstable

Main sequence stars remain stable, or in equilibrium, for a very long time, but not forever. This period may last from millions to billions of years. Eventually, their cores run out of hydrogen fuel. When this happens, pressure decreases and gravity takes over, and becomes the stronger force. At this point, the star is approaching the end of the main sequence stage.

Do Stars Fall?

Have you ever seen light blaze across the night sky and then disappear? It looks like a star has fallen from outer space. Many people call it a falling or shooting star. But don't be fooled. Stars don't fall!

What are those streaks of light in the sky?

The light streaking across the sky is really caused by meteoroids— bits of dust or rock in space. Meteoroids burn as they pass through Earth's atmosphere, creating stunning trails of light called meteors.

The brightest meteors are called fireballs.

Why are meteors different colors?

Meteoroids create friction as they hurl through Earth's atmosphere. Friction heats the meteoroids and makes the air around them glow. Their light trails may appear white or in color, depending on how elements in the dust or rock react to gases in Earth's atmosphere.

Why do meteor showers appear around the same time every year?

If you see a lot of light trails on the same night, or over several nights, it might be a meteor shower. Meteor showers occur thanks to icy, rocky objects called comets. As comets **orbit** the sun, they leave dust and rock behind. Annual meteor showers occur when Earth passes through that dust and rock.

Debris from Comet Swift-Tuttle creates the Perseid meteor shower, which is visible for parts of July and August.

Supernovas are the biggest, most violent explosions in the universe. Their brightness can take months to fade.

CHAPTER 4

The Final Stages

Up to this point, all stars have gone through the same stages. They have formed in nebulas, become protostars, and then main sequence stars. When a main sequence star runs out of hydrogen fuel in its core, the helium that was created is still there. The core becomes hot enough for helium to fuse, creating carbon. This nuclear reaction creates a huge amount of heat and light, causing the star to grow and release more energy. As it grows, it becomes brighter. Eventually, the star cools and glows red. Depending on its mass, it will turn into a red giant or a red supergiant.

The End of a Red Giant

When a red giant runs out of helium, it stops producing carbon. Its core then collapses and releases its outer layers of gas. It is now a planetary nebula. The nebula blows apart, leaving a hot and dim Earth-size star called a white dwarf. After billions of years, a white dwarf will cool down, leaving a star called a black dwarf.

Timeline of the Universe

13.7 BILLION YEARS AGO:
The universe forms in an event called the big bang.

ONE SECOND AFTER THE BIG BANG
Neutrons, protons, electrons, and other tiny particles form.

10 SECONDS TO 20 MINUTES AFTER THE BIG BANG
The first hydrogen and helium atoms form.

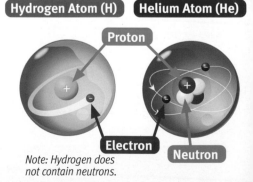

Hydrogen Atom (H)　　Helium Atom (He)

Proton

Electron　　Neutron

Note: Hydrogen does not contain neutrons.

The End of a Red Supergiant

When a red supergiant runs out of helium and stops creating carbon and other heavier elements, its core also collapses. This causes a violent explosion known as a supernova. After the explosion, some supergiants become neutron stars—dense objects with all their matter squeezed into a radius of about 12 miles (19 km). The most massive collapsing stars form black holes—invisible objects in space with gravity so powerful, light cannot escape.

150 MILLION TO 200 MILLION YEARS AFTER THE BIG BANG
The first stars form, mostly from hydrogen and helium.

ONE BILLION YEARS AFTER THE BIG BANG
Stars form the first galaxies.

NINE BILLION YEARS AFTER THE BIG BANG (4.6 MILLION YEARS AGO)
Our sun forms.

Life Cycle of a Star

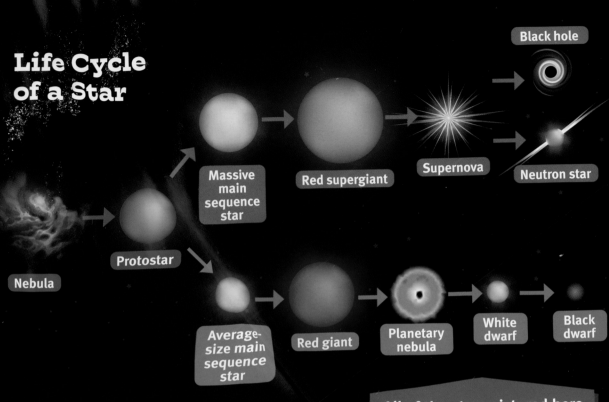

Nebula

Protostar

Massive main sequence star

Red supergiant

Supernova

Neutron star

Black hole

Average-size main sequence star

Red giant

Planetary nebula

White dwarf

Black dwarf

All of the stars pictured here exist now, except one. Black dwarfs still need trillions of years more to form.

Going Through a Cycle

For millions of years after the big bang, no stars existed. Today, nebulas in our universe create billions of stars each year. Every star is born in a nebula. And every star moves through a life cycle that will lead to an end stage as a black dwarf, neutron star, or black hole.

Using Up Fuel Fast

Glowing brightly, red supergiants have more fuel to fuse than red giants. So you might expect them to live longer. But they don't. Why? The answer lies in that invisible force—gravity. Massive stars have stronger gravity than smaller stars. This makes their cores grow hotter, which causes nuclear fusion to occur faster. The result? Red supergiants use up their fuel much faster than red giants.

VY Canis Majoris

Sun

Red supergiant star VY Canis Majoris is one of the largest known stars in the universe.

Astronomers use different types of telescopes to study stars and other objects in our universe.

Seeing in the Dark

Stars are hard to see because they are so far away. Scientists use telescopes to bring views of the stars closer and explore the universe. Some telescopes are based on Earth. Others are placed in space. A radio telescope, used to detect radio waves, is one kind of telescope. Another type, called an optical telescope, does two things to make objects easier to see: It collects light and makes objects appear larger.

The Largest Telescope on Land

The Gran Telescopio Canarias (GTC) is the largest telescope on Earth. This optical telescope uses mirrors to focus light and make objects appear closer. The GTC is located on a mountaintop in the Canary Islands of Spain. Its dome stands 147 feet (45 meters) tall, towering over the other telescopes in the area. The GTC is so powerful it can detect the flame of a candle from 20,000 miles (32,187 km) away!

It took nine years to build the GTC telescope.

Light Pollution

Have you ever looked up on a clear night but found it was not dark enough to see many stars? When artificial light makes the night sky difficult to view, it's called light pollution. Unwanted light makes it hard for astronomers to do their jobs and for people to enjoy the natural night sky. Some communities work to reduce light pollution. They ask people to turn unneeded lights off and close curtains to stop light from seeping outdoors. By taking these simple steps, you can help lower light pollution, too!

The bright lights of New York City make it hard to see stars in the night sky.

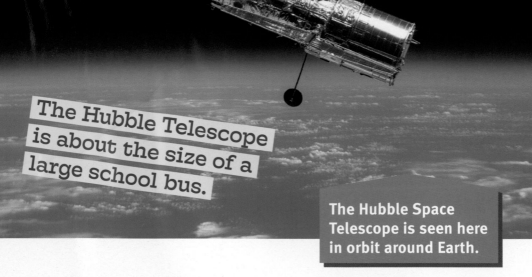

The Hubble Telescope is about the size of a large school bus.

The Hubble Space Telescope is seen here in orbit around Earth.

The Hubble Space Telescope

In 1990 the Hubble Space Telescope (HST) was launched into orbit around Earth. It was the first optical telescope based in space. The HST orbits Earth at about 17,000 miles per hour (27,300 kilometers per hour), taking pictures of stars, planets, and galaxies. It weighs about the same as two male African elephants and runs completely on solar power—energy from the sun!

Radio Telescopes

Radio telescopes look like huge satellite dishes. They have antennas that can detect radio waves from stars, planets, galaxies, and other objects in space. This helps astronomers learn about the structure and motion of space objects, along with the elements they contain. Actually, telescopes help astronomers learn more about the universe, from galaxies to the amazing life cycles of stars. In astronomy, the universe is the laboratory!

An array, or group, of radio telescopes

Studying Star Brightness

A variety of stars shine in outer space.

The Hertzsprung-Russell diagram compares a star's surface temperature and its luminosity. It was created in the early 1900s by two astronomers, Ejnar Hertzsprung and Henry Norris Russell. Placing stars on the diagram helps scientists identify which star types are most common.